MW00678912

A NEW
APPROACH
TO GLOBAL
WARMING

A NEW
APPROACH
TO GLOBAL
WARMING
The Unacknowledged Cause

FRED LOVE

TATE PUBLISHING
AND ENTERPRISES, LLC

A New Approach to Global Warming
Copyright © 2016 by Fred Love. All rights reserved.

No part of this publication may be reproduced, stored in a retrieval system or transmitted in any way by any means, electronic, mechanical, photocopy, recording or otherwise without the prior permission of the author except as provided by USA copyright law.

This book is designed to provide accurate and authoritative information with regard to the subject matter covered. This information is given with the understanding that neither the author nor Tate Publishing, LLC is engaged in rendering legal, professional advice. Since the details of your situation are fact dependent, you should additionally seek the services of a competent professional.

The opinions expressed by the author are not necessarily those of Tate Publishing, LLC.

Published by Tate Publishing & Enterprises, LLC
127 E. Trade Center Terrace | Mustang, Oklahoma 73064 USA
1.888.361.9473 | www.tatepublishing.com

Tate Publishing is committed to excellence in the publishing industry. The company reflects the philosophy established by the founders, based on Psalm 68:11,
"The Lord gave the word and great was the company of those who published it."

Book design copyright © 2016 by Tate Publishing, LLC. All rights reserved.
Cover design by Lirey Blanco
Interior design by Gram Telen

Published in the United States of America

ISBN: 978-1-68333-126-1
Science / Global Warming & Climate Change
16.06.06

Contents

The New Approach

Back in the 1940s, my dad worked for the carbon factory in Clarksburg, West Virginia. While he was there, they were having trouble firing up the main furnaces. The process they used was to fire up the first one and work their way back. He was a very innovative young man and suggested to his supervisors that it would be much easier and more efficient to fire them up from back to front. It worked so well they used the procedure for many years. I mention this because he taught me to look at things from a different angle or perspective.

I remember one time I mentioned an idea to my dad for a new bottle cap. It was very similar to the present-day twist-off cap consisting of aluminum with slots on the side that part when turned. My idea was to make the top dome like such so one could pop or smack it with their hand to open it. He said I looked at it from a different view. Of course, it was not practical because of shipping. You would have a lot of bottles pop open all due to bouncing around in the delivery trucks.

You ask, "What has this got to do with climate change or global warming?" It's because I am looking at it from an entirely different view. Some people seem to see climate change as the drive by media. If it is hot outside, the planet is warming. If it is cold, we are entering a new ice age.

Many scientists ignore real facts because they will lose funding or prestige. I was recently at a global warming discussion. The host was using it to gain notoriety on the web and wanted to be associated with a scientist involved with climate change in order to create prestige for more donations from uninformed people who have sincere intentions. He was totally out of sync with his explanations on global warming.

Oil and natural gas companies will try to dismiss my theory and in every way try to disprove it. I welcome this wholeheartedly because if I am wrong, it would be a great relief and excellent news for mankind, especially for our grandchildren and great-grandchildren.

I worked on the Contemporary Hotel at Disney World in Orlando, Florida, when it was under construction by United States Steel Corporation. I noticed that almost everything was made out of steel, even little things that could have been made with lesser and easier materials. Their mind-set was steel, and it was difficult for them to think otherwise.

Not that this was wrong. After all, they were in the business of selling steel; and the more things they made out of steel, the more money they would make.

Again you might ask, "What does this have to do with climate change?" I'm trying to show that people and companies get locked into a particular mind-set or process that is very hard or difficult to avoid. There is always another approach to a situation or problem.

So let's get started with a new approach to what really is causing global warming. I prefer to use the phrase "global warming" rather than "climate change" because it actually fits my theory better. However, it must be used carefully, which I will explain later in this book.

It's the core. There I said it. Now let's prove it. First let's give some reasons. My theory is basically that global warming is caused by heat seeping up through the interior of the Earth. This is caused mainly by the removal of oil and natural gas that is protecting us from the extreme heat from the core. In order to explain how I came to this conclusion, let's examine my reasoning.

The Earth is cooled much like an engine. It is cooled by air, water, and the block that holds it together, much as the crust of the Earth does. Engine oil protects the engine from overheating. Water also cools the engine as it flows through the radiator. The air cools the engine as it helps cool the water and flows over the engine, just as the wind covers the Earth's surface.

When any one of these is removed, the engine will overheat. The Earth works in a similar way. When the wind stops blowing, it gets hotter. When it does not rain, it causes a drought, and it gets hot. The Earth's core is a molten, hot material believed to be mostly iron (Fig. 1). Some scientists believe the core to be as hot as the surface of the Sun. This molten hot material is very close to the surface, such as in Dallol, Ethiopia, where we can see hot lava bubbling up to the surface.

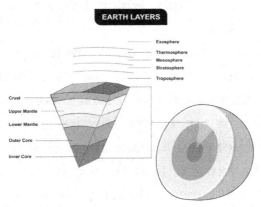

Figure No. 1: Make up of the Earth's core

We are protected from this extreme heat by the water that covers over 70 percent of the Earth, the land that comprises approximately 30 percent, and the air that covers the entire surface. That leaves us with oil and natural gas.

The crux of my theory is that this protection of oil and natural gas from the interior of the Earth is being removed at an alarming rate. Oil is being extracted in excess of ninety million barrels of oil a day as of 2009 along with 515 billion cubic meters of natural gas. This does not account for minerals deep in the Earth such as copper, iron, silver, gold, etc. The removal of oil and natural gas is leaving vast areas of Earth's crust unprotected from the heat generated by the core of the Earth. One might say it is beginning to look like Swiss cheese in places like the Middle East, the Gulf of Mexico region, Siberia, Alaska, and the North Sea. You could make a case that it is all over the world. This depletion of oil and natural gas in these areas is causing heat to rise to the surface. This seepage has even been felt physically in warmer climates, but the seepage from the core in the extreme northern parts of the world has seen ice melt at an alarming rate.

This theory shows a more plausible explanation than greenhouse gasses. It can also explain the increase in the frequency and sizes of earthquakes and drought in such places as California, Texas, and Oklahoma. This theory can also explain why it is colder or hotter in certain areas of the world as well as explain the unusual weather patterns we are experiencing and the melting of snow and ice in Greenland and Iceland, which I will prove in a later chapter (Fig. 2).

Figure No. 2: Rasmussen Glacier, East Greenland

Let's consider some more examples such as this. The Sun is approximately 94,000,000 miles (151,000,000 kilometers) away. Yet we ignore the Earth's core that some scientists believe to be hot as the Sun. Only twenty-one miles (thirty-five kilometers) of the Earth's crust separates us from the mantle's extreme heat, believed to be composed of molten metals like what we see when there is a volcanic eruption. Twenty-one miles is much closer compared to ninety-four million miles.

Practically everyone was glued to the television when Chilean miners were trapped underground in 2010. The temperature where the miners were was approximately 120 degrees. On top where the rescue workers and families where it was around 40 to 50 degrees. As the miners were brought to the surface, one could see steam exiting from the rescue hole. They were quickly wrapped in blankets and coats to protect them from the extreme change in temperature. This could be an example of what is constantly occurring when oil and natural gas are brought up to the surface.

Another example: What if you drill small holes in your house or business for a period of time. Eventually, the heat or cold would soon escape to a point it would no longer be comfortable, let alone be advantageous to the cost of your heating and air conditioning bill.

Another visual example is the use of a pressure cooker (Fig. 3). A pressure cooker is used to cook food under pressure. It is similar to the Earth, which is much like a pressure cooker churning with hot gasses and cooking its material under tremendous pressures. A pressure cooker releases its pressure by a weighted pressure regulator on top and maintains the pressure inside the cooker. The regulator releases pressure during the operation, which is very similar to how a steam engine works. It actually levitates above the nozzle, allowing excess steam to escape. Everyone has seen pictures of oil wells shooting oil into the air. The well has penetrated the pressure cooker under the ground, releasing excess heat resulting from the cooking of oil and natural gas over millennia inside the Earth. This is caused by pressure from trapped oil and natural gas, which has been pushing up on the Earth's crust. The crust, oil, and natural gas are protecting them from the extreme heat from the Earth's core. We are interrupting the natural barrier protecting us from the extreme heat found a mere twenty-one miles under our feet. Volcanoes act as a safety valve, similar to the safety valve on a pressure cooker. There is a backup system in a pressure cooker used as a release mechanism, which is a hole in the lid made of a rubber grommet with a metal insert at the center or alloy plug that releases pressure very quickly if the pressure regulator becomes clogged by food or seeds. Volcanoes erupt when the pressure becomes too high. There are regulators (such as Old Faithful in Yellowstone Park) releasing pressure all over the world. Man now is adding millions of such releasing holes by drilling into the Earth's pressure cooker, thereby releasing heat into the atmosphere.

Figure No. 3: Pressure Cooker

Removing this protective insulation is causing and will cause serious climatic and global problems now and into the future. Once this protective insulation is removed, it cannot be replaced. One could very easily argue that pollution above the surface caused by automobiles, manufacturing plants, fires, planes, etc. can be corrected in the next fifty or so years, but oil and natural gas are not renewable and cannot be replaced.

In the following chapters, I will present many effects caused by this extraction of the lifeblood of the Earth, and this will or could explain what is occurring. I am not declaring anything as absolute, but I will wait for the academic community to prove or disprove my theories. Some will be conjecture while others can be easily verified. One person or group will not be able to comprehend the enormous task. This will require covering all the areas that this new approach to global warming brings to the forefront. I am only asking for truthful scientific evaluation without prejudice, station, or monetary value. I cannot express how crucial this is to the future generations of our planet.

"Is It Hot or Cold Outside?"

Is it hot outside? If it is hot, how hot is it? How long will it be this hot? Will it turn cold? If it turns cold, how cold will it get? If it's going to get cold, how long will it stay cold? These are questions we want to be answered. However, it is hard to give an exact answer. Some will never be answered. Others, as the following will show, will not need a conclusive answer. When I started researching global warming around twenty years ago, I estimated that record-breaking temperatures and extreme weather events would become commonplace at around 2025. At the time, I did not know or anticipate that the United States would become the world's largest producer of oil and natural gas. But just because the United States is the major oil and natural gas producer in the world does not let other nations off the hook such as Russia, Saudia Arabia, the Middle Eastern countries, and even smaller producers like Mexico, India, and China, just to mention a few. Needless to say, I was wrong. Record temperatures are being broken every day.

Has anyone noticed that temperatures are higher around oil- and natural-gas-producing regions? We cannot determine how hot our planet will get because of several factors. The primary factor is when and how much more oil and natural gas as well as geothermal energy is extracted from the Earth. Another is the location and timing of the continuous removal of oil and natural gas out of certain areas over the world.

Figure 4a: Mount St. Helens Blow-off

At the time of the writing of this book, a new glacier has been discovered of all places in a volcano called Mount St. Helens located in Washington state (Fig. 4a). When it erupted, it blew 1,300 feet off the top of the mountain on May 18, 1980, and killed fifty-seven people and destroyed twelve glaciers that existed on top of Mount St. Helens at the time. This is a perfectly natural example of what is going to happen to the future of our planet. I will explain why this is so critical.

Scientists estimate that around four and half million years ago, the crust of the Earth was about four and a half miles thick. I think they put the half in this estimate to make it appear more accurate than it is. Today it is an average of twenty-one to twenty-four miles thick. This means that the mantle, which is hot lava, is losing its heat and increasing the thickness of the Earth's crust over time. However, man's intervention in this process by extracting oil and natural gas is causing this outer layer to enlarge at a greater rate than it would have done so at a more natural pace. This process would have taken thousands of years instead of decades to occur as it is today.

When volcanoes erupt, they release an enormous amount of heat held under pressure that has been brought up from deep in the Earth's core. Similarly, referring to the example of the pressure cooker, if the release valve was made bigger, thereby letting greater amounts of pressure to escape, it would take a longer time to build up its pressure depending on the proximity of its heat source.

Figure No. 4b: Mount St. Helens

Scientists have found a glacier developing on Mount St. Helens, and it is expanding but has slowed down due to hot summers (Fig. 4b). They say it is being shielded by a nine-hundred-foot-high dome pushed up by hot magma inside the crater, thus shielding the glacier from the Sun. Does this make any sense? The answer is yes and no.

First of all, the Sun does not shine all the time. There is a nighttime. It is also seasonal in Washington state, meaning there are certain times of the year that are colder and certain periods that are warmer. What I am trying to say is that the heat from under this new glacier has been released by volcanic eruption into the atmosphere, thereby enabling the glacier to form not because of shade from the Sun but because of the absence of heat from below, which is protected by a thicker layer of earth. Keep in mind that Mt. St. Helens is still 8,363 feet (2,549 meters) tall. Before its eruption, it was 9,677 feet (2,950 meters). There are still more than a mile of the Earth's crust above the surface, and there is still twenty-four to twenty-five miles from the average distance to the mantle. This is what I am trying to show. The thickness of the Earth is a better protector of heat than anything above the Earth's surface. Therefore, as the crust gets thicker by losing its heat, the colder it is going to get. How quickly heat is released from beneath the ground will determine how cold it will get. The process is not determined by heat from the Sun, as activists would want you to believe.

Temperature measures the intensity of something, but heat measures an amount of something. Heat is energy being transferred from one object to another because of temperature differences alone. But the glacier is being insulated by the protective crust underneath. Therefore, transfer of heat cannot be achieved. When a second slower volcanic eruption in 2004 occurred, a nine-hundred-foot dome was formed. That dome was thought to be shielding the glacier from the Sun. It was also thought the hot magma from the the explosion would also melt the glacier, but this did not happen because there was no transfer of heat between the hot magma and the ice glacier. There were some slowing of the expansion of the glacier, not from the Sun, but from a transfer of a small amount of heat from the hot magma.

The Sun shines on all glaciers throughout the world. For example, the Sun shines just as long and intensely in the Antarctic, which is growing in size, as it does in the Artic, which is decreasing in size. What is the difference? You got it! It's oil and natural gas.

The conclusion of this is that as heat is transferred to the surface, the crust is growing thicker. Therefore, our planet will over time become colder, especially in the northern latitudes. We have inadvertently and prematurely started the new ice age sooner than it would have been. It's going to get cold outside.

We cannot see this happening. It's very subtle. Astronauts like Scott Kelly on the International Space Station has been taking beautiful pictures of our planet from space. As we look at these beautiful snapshots of the Earth, we cannot see what is happening inside. The planet has so many wonders of beauty for the eye to behold, yet inside it is groaning under the pressure of man's relentless pursuit of energy. If we do not heed the warning signs, this beauty will be transformed into just another lifeless ball in the universe (Fig.5).

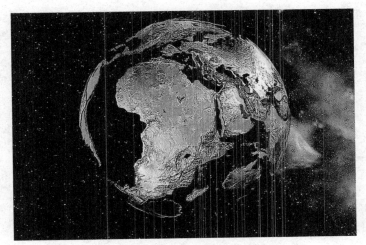

Figure No. 5: Is this our future?

The Earth Groans

Oil and natural gas are being held under immense pressure within the Earth's crust caused by the compression of the tremendous weight of rocks and water. When released to the surface, oil and natural gas leave large caverns or spaces that could collapse, resulting in tremors and possible cartographic earthquakes.

An example of this would be placing a balloon on a table then putting pressure onto the balloon with one hand while using a pen to puncture the balloon, causing the air to escape out of it. The result would cause your hand to smack onto the table. Wouldn't rock deep in the ground be susceptible to the same thing? When a drill hits oil and natural gas, it releases this enormous amount of pressure it contains. Wouldn't it cause the earth inside the cavern to collapse under the immense pressures from above, thereby creating tremors or even earthquakes?

In Florida, this happens quite often with underground water caverns. They are called sinkholes, and they are very close to the surface (Fig. 6). When they collapse, they sometimes devour homes, cars, roads, and businesses.

Figure No. 6: Sink Hole - Winter Park, Florida May 1981

The US Geological Survey and the Oklahoma Geological Survey have confirmed a connection between the recent oil and natural gas boom taking place in the United States with the rise in seismic activity in Arkansas, Colorado, Ohio, Texas, and Oklahoma. They say it is because of the injection of wastewater deep into the ground. They call this system of extracting oil and natural gas as *fracking*; we will cover this process later in the book.

Some of these quakes have been registering a magnitude of 5.6. I believe it is a direct result of the removal of oil and natural gas out of the ground. It is leaving huge canyons to cave in, causing tremors. I would venture to say that the wastewater is increasing these tremors, but it is not the original cause.

I understand that oil and natural gas are the economic life of these states, but the future of our generations to come should be of greater concern. It is tough to do this, but it should be our utmost priority to slow this down in an extreme way.

Realize that this will be hard and challenging to accept. However, logic should prevail until we know for sure what these tremors are doing or how they will do damage to our world before we cross a line of no return or move in a direction that can never be corrected.

I remember when I was very young I wanted to make space under our house larger so I could use it as a play area. So I started digging out a wall made of clay. Of course, when my dad came home, he explained to me that I could not make it any bigger because I was removing the support from underneath the house, and it could eventually collapse. This same thing is happening underground due to the removal of oil and natural gas. Let's ask a simple question: Why do miners shore up mine chaffs? If they were not concerned that a cave-in would occur, why do they waste all that time, trouble, and expense?

An example is the 7.9-magnitude quake in China's Sichuan province in 2008 that is believed to have been caused by water behind a dam pressing down on a fault line. The disaster left 4.8 million people homeless. Here is the problem with men thinking in one direction and ignoring what should have been easily analyzed in another direction.

"Do not build anything that puts undue pressure on a fault line." The federal government has stopped any drilling near a fault line after tremors were felt while drilling close to a known fault. These tremors have occurred in Oklahoma.

There are twelve tectonic plates around the world (Fig. 7). They move on liquid that can be molten lava, or it can be oil and natural gas. Remove this oil and natural gas, and the plates can stop, fall, or cause other liquids to evade. In either case, plate movement will occur. If it stops, other plates that are still moving will collide with it. If the plate drops, other plates could rise by its deflection. In any case, there is going to be an increase in earthquakes throughout the world, due to man's intrusion into the natural process of our planet.

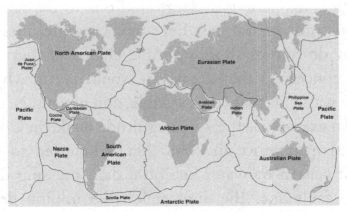

Figure No. 7: Location of Tectonic Plates

Figure No. 8: Alaskan pipe line with heat pipes

We are extracting ninety billion barrels of oil out of the earth every year or 246,000 barrels per day; this does not include natural gas that is around 154 billion cube feet per year. When oil and natural gas are extracted out of the ground, it is not only pressurized but it is also extremely hot. The oil coming out of Prudhoe Bay in Alaska has a temperature of up to 145 degrees Fahrenheit when it exists from the oil wells. They even had to use heat pipes to dissipate the heat (Fig. 8). Without such pipes, heat from the oil would cause the permafrost to melt; and in turn, the pipe supports would sink. To prevent such a disaster, more than 12,400 heat pipes were mounted on top of the pipelines' vertical supports. These are used to keep the permafrost frozen and intact by conducting heat from the supports out in the air and not in the ground. When the oil is moving throughout these pipes, it must be kept at its pour point, determined by the amount of paraffin wax it contains. This temperature can be as high as 48.2 degrees Fahrenheit (32 degrees Celsius). It is very warm compared to the outside temperature. To sustain this pour point throughout the pipeline, they not only need to protect the permafrost but they also need to keep the oil flowing through the pipe. To achieve this, they use several means to do so. One way, as in the Alaska pipeline, is to suspend the pipes above ground and let the heat be released out into the air. The pipeline is also insulated to keep the heat from escaping. When the pipeline is underground, it uses the ground to insulate the pipe further. However, the heat from these pipes is still released into the ground and the air. The fact is that there are over 800 miles (1,287 kilometers) of pipe that are 48 inches (122 centimeters) in diameter. There are also hundreds of miles of feeder pipelines crossing Alaska, radiating heat from its source deep in the ground and, for that matter, from thousands of miles in other areas across the world. Just to mention a few, some of them are in Russia, Canada, Europe, and the United States. All these pipelines are releasing heat from their source deep in the ground or oceans into the atmosphere and ground far away from its original beginnings. The United States has over 1.2 million miles of pipe alone.

Let's look at an example of how radiated heat is being used today to understand what is happening with hot oil moving through pipelines. Many homes are putting heated floors composed of concrete, stone, tile, or carpet. This form of flooring is being heated from below one's feet in their home. They have been using this system for many years, even using it to keep snow and ice from forming on driveways.

Figure No. 9: Oil Tanker Convoy

We could logically assume that the same process is being created by thousands of miles of pipelines being buried all over the world—pipelines that are used to transport oil to refineries and tankers, which leads to another question about radiant heat from the oil. Are ships carrying an enormous amount of oil across our oceans 24-7 putting heat that is not natural into our oceans and streams throughout the world (Fig.9)? Are they causing the warm tropical currents to be heated more than they were designed to be? Is this causing an increase in the temperture of the the thermohaline circulation system where the ocean waters are being heated in the tropics? Could these waters be enhanced by the oil barges that must keep their cargo of oil at a certain temperature called the *pour point* in order to load and unload their cargo? I will explain this further in the book.

I want to state that this is a subtle process that is very slowly creeping into our oceans along with the exhaust from millions of ships of all kinds. One cannot point to a single oil carrier and state its heat impact on this planet. In reality, it is insignificant by itself but gathered together with thousands of other ocean-going vessels, it has an enormous effect on warming up this planet called Earth.

When I was young, we lived in a two-story house that was being heated by a coal furnace in Clarksburg, West Virginia. I can remember playing in the coal bin against my mother's wishes. Trust me, I could not hide what I had done; I looked much like a coal miner coming out of the mines. My dad bought a coal mine once in order to save on heating cost. He loaded two truckloads of coal and delivered it to our house. He said it was the hardest day's work he had ever done. Needleless to say, the mine was sold very quickly after that. I was much like the kid in the movie *Home Alone* when I first saw the inside of a coal-burning furnace with the hot coals burning like the lava flow from a volcano.

The heat from the coal furnace was then transferred through a duct system into three rooms on the first floor; from there, the heat moved upward through vents to the second floor. I use this example to show how heat is radiated up from a lower place. It is the same way heat radiates from oil and natural gas wells, even mineral mines that go deep into the Earth.

At present, the Alaska pipeline is only carrying one-third of its capacity, which starts another problem. It is called revenue for the government. In Alaska, nearly 90 percent of their income into the government coffers comes from oil and natural gas. Lack of revenue causes government politicians to relax rules or give incentives in order to get more money. At this time, the governor of Alaska is renewing his effort to get the federal government to open up the Arctic National Wildlife Refuge to oil and natural gas drilling. They consider this huge area of land as easy pickings for drilling. According to my theory, it would put the final nail in the coffin, so to speak. More pipelines would be needed. However, before that, the amount of protection from the Earth's interior heat would be very quickly eliminated. There is an estimated fifteen billion barrels of oil in the Chukchi lease area alone. When governments have the loss of revenue and loss of jobs at the same time, much heat is put on those in charge, but they will survive. However, the greater heat will be from the heat escaping from all the new oil wells that will be drilled, which will not regain their insulation that is protecting us.

The point I am trying to make is "Why is it so hot so far north and releasing so much heat out of the ground?" This pressure and heat can sometimes be released very quickly. When the two are eliminated— pressure and heat—it causes their former location underground to contract, thereby causing tremors. These tremors could in turn cause devastating earthquakes, depending on how much pressure and heat is involved. It could have been the reason the Great Alaska Earthquake occurred in 1964. And it could very easily happen again if oil and natural gas are allowed to be pumped out of the Arctic National Wildlife Refuge.

Alaska's first oil wells were in 1898, striking small amounts of oil. These lasted until massive amounts of oil and natural gas were found in Texas and Oklahoma and flooded the market with cheap oil, causing it to be too expensive to extract oil out of the ground in Alaska. Most of the oil activity in Alaska stopped at this time, but in 1957, a large deposit of oil was found on the Kenai Peninsula called the Swanson River oil field. In 1959, they found a major natural gas field near the Swanson River oil field. Millions of barrels of oil and natural gas started to be removed every day, releasing heat and depressurizing the Earth's surface, thereby upsetting the balance of the natural order. It is no wonder that the Earth is groaning under the act of taking out the very lifeblood of the Earth and transferring it into man-made veins of pipelines all over the Earth.

Recently, the federal government renamed Mount McKinley to Mount Denali (Fig. 10). They found that the mountain was several feet lower than originally thought. The modern instruments used were considered more accurate than the instruments used earlier. However, could these instruments have been wrong? It was not because of faulty instruments but because of the removal of the support of oil and natural gas under the crust of the Earth, which is very much like how it was when I removed the clay from under our home. Logic tells us and experiments show that the immediate vicinity does not have to be real close to a structure to cause damage.

Figure No. 10: Mount Denali, Alaska

At this point, I would like to mention something about the earthquake in Nepal on April 25, 2015. The cause was due to the collision of the Eurasian tectonic plate and the Indian tectonic plate. There are some groups of scientists using satellite imagery before and after the earthquake that have determined that Mount Everest has decreased in height due to the earthquake. A very small decrease of 2.5 centimeters is not a very large decrease in height. I will give this scenario of why this happened instead of the more common event of causing mountains to rise higher due to the tremendous collision of these plates. I call this the "seesaw effect." It works like this: when the Indian Plate collides with the Eurasian Plate, it puts incredible pressure upward into the Eurasian Plate, causing the upheaval of the ground, which over time created Mount Everest. However, this time, man has come along and interjected himself in the natural course of events by the extraction of large volumes of oil and natural gas out of the Earth, not near to these enormous colliding tectonic plates but on the opposite end of the Eurasian Plate in Russia. Russia has been pumping oil and natural gas out of the Siberia for almost forty years at the rate of millions of barrels per day. The result of removing this large amount of oil and natural gas out of the Earth leaves huge caverns, which results in leaving the area unsupported. I believe that when the Indian Plate collided with the

Eurasian Plate, the Eurasian Plate seesawed down in the northern part. And then, because of the weight of Mount Everest, the plate came back down and rested near its original height before the Gorkha, Nepal earthquake occurred. Please understand that we are talking about a slight rise in height, possibly as little as 0.5 centimeters. This rocking of the Eurasian Plate occurred over thousands of miles is very slight; however, it is a possibility. I mention this because if this is true, then Nepal will have another earthquake sooner than is natural to occur all due to man's interferences in the natural progression of our planet.

I am going to mention another unexplained phenomenon put out by *National Geographic*: a new research study that shows that the North Pole is changing its position at an alarming rate of forty miles per year toward Russia. It was over Northern Canada but has migrated more toward Russia. As the research states, scientists believes that changes deep in the Earth's molten core are the culprit. Researchers have detected a disturbance on the surface of the core that is creating what they call a "magnetic plume," which they believe is causing this shifting of the magnetic pole. They are not sure what has caused these disturbances. Could it be caused by the opening up of huge caverns near the molten mantle that is letting the core ooze through the protective rock deep in the Earth? The scientist has made 3-D images that show what is believed to be a geothermal body in other places that look like plumes. Could these geothermal-heated plumes be a heavier, more solid magma that is creeping up toward the surface and causing the disruption in the location of the North Pole?

At this point, I want to suggest another possible event that is occurring. It is the transfer of weight from one area of the globe to another place. An example of this is the Three Gorges Dam in China reservoir that contains about 9.43 cubic miles (39.3 cubic kilometers) of water. The water weighs more than forty-two billion tons (thirty-nine trillion kilograms). NASA scientists calculate that the weight of water in this one area has caused the Earth's mass to shift, causing the rotation of the Earth to take longer and thus slowing its rotation. Due to this one area, it is estimated that the length of the day is increasing by 0.06 microseconds, making the Earth slightly more round in the middle and flat on top. It would shift the pole position by about 0.8 inches (two centimeters). It is already happening in Russia, Canada, Brazil, and other midlatitude countries for over forty years. Okay, what does this have to do with oil? An example would be the Trans-Alaska Pipeline. I could also use the Russian pipeline, the North Sea oil wells, the Canada oil sands, and much more. Think about this: the Alaskan Pipeline has been extracting oil out of Perdue Sound since June 20, 1977. Almost seventeen billion barrels of oil have moved through this one pipeline alone. A barrel of crude oil weighs between 125 to 140 pounds, depending on whether it is light crude or heavy crude oil. When we multiply seventeen billion barrels by a conservative number of approximately 130 pounds per barrel, we get over two trillion pounds. For comparison, dividing this by two thousand pounds per ton, we get one billion tons of crude oil transferred to the lower forty-eight. It is not quite the equivalent to the weight of the Three Gorges Dam, but this is only one pipeline. A pipeline does not have to be long to be detrimental in transferring weight to other regions of the world. I am just saying that other things are contributing to the cause of increasing the mass of the middle section of the planet besides one location.

The weight shift could be the reason behind the movement of the North Pole and also the vacant space created below the surface due to the enormous amount of oil being extracted. It should also be noted that this is not occurring at the South Pole. The difference in the North and the South Pole is the extraction of oil and natural gas.

During the writing of this book, there has been much talk about the weather phenomenon called *El Niño*. El Niño is an unusually warm ocean temperature in the Equatorial Pacific region. Scientists say moderate trade winds in the Pacific cause this phenomenon by stalling the upwelling of colder, nutrient-rich water to the surface. Even in this weather system, we are looking in the wrong direction. The Sun has not changed its course. The Earth has not changed its course, and neither has the moon. So what has changed? A new approach would be to consider the tectonic plates. They move. They change. The Pacific Plate, which is the largest tectonic plate, is moving toward the northwest. Around its boundary on the west, north, and east is an area called the "Ring of Fire" (Fig.11). The Ring of Fire is where a large number of earthquakes and volcanoes occur. There are over 452 volcanoes there, and it is estimated to have over 75 percent of the world's active and dormant volcanoes. In the middle of this plate is a hot spot called the Hawaiian hot spot. There is another in the Galapagos Islands and five other smaller hot spots to the south.

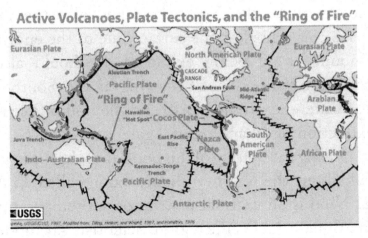

Figure No. 11: "The Ring of Fire"

I would like to interject another view of the string of Hawaiian Islands. Scientists refer to the islands as curving to the Earth's surface, but using my earlier example of a pressure cooker, you will notice that the lid is domed to let pressure escape at the top. Tectonic plates are designed the same way. They are domelike, so the curvature of the islands are developed underneath by heat radiating to the middle or highest point of the tectonic plate. I am pointing out that these islands are controlled under the crust, not above. Also using the pressure cooker example in comparing it to the Ring of Fire, the pressure cooker would be leaking its heat around the seal. It should be noticed that there is a string of hot spots between Australia and South America just south of the equator. Has anyone looked into the possibility these hot spots are releasing more heat than normal?

Concerning the recent Australian drought, three volcanoes have recently been found in Southeastern Australia. It may be too soon to determine, but could these volcanoes be responsible for the drought in Australia? One volcano spits out ash after magma is quickly cooled. Why did they just find these volcanoes? Could these volcanoes have erupted to a point that they could be detected by releasing more heat than usual? Did this new heat source help cause the Australian drought? Australia is located in the middle of the Indo-Australian Plate, similar as to how Hawaii is located in the middle of the Pacific Plate. Heat migrates to these points, and when you remove the protection of the insulating qualities of oil and natural gas, no wonder it gets hot. Australia extracts over 150 billion barrels of oil per year and over 5.6 trillion cubic feet of natural gas.

The Indo-Australian Plate and the South American Plate are moving east; the Nazca Plate is also moving east. As stated earlier, the Pacific Plate is moving northwest. These movements cause a greater gap in the plates that causes more heat to rise in the eastern part. The cause can be linked to the earthquake in Nepal and the quakes in China, Japan, and all along the western boundary of the Pacific Plate. When the Indian and Asian Plates collided and closed the distance between them, it enabled other plates to move easier, consequentially allowing the Pacific Plate to open in the east a larger gap that heats up El Niño. Bear with me here. The Pacific Plate is estimated to move 2.76 inches (7 centimeters) per year. One might say that a movement of only 2.76 inches or 7 centimeters per year is so small and might question why it matters. If we approach this differently, we come up with a much bigger number. The Ring of Fire along North and South America has many deep trenches. For example, the Peru-Chile

Trench, also known as the Atacama Trench, is about 100 miles (160 kilometers) off the coast of Peru and many others along North and South America. These trenches have been releasing enormous amounts of heat naturally for thousands of years. Consider also that according to NOAA temperature maps, the increase in temperature is highest in the middle of the Pacific Ocean, not the warm air coming from the drought on the Australian continent. The heat intensity map of El Niño shows the heat is most intensive north of Australia and out in the middle of the Pacific Ocean.

Another reason to show the increasing heat forming El Niño is the removal of oil and natural gas from Chile, Peru, Ecuador, Colombia, Guatemala, and Mexico, which are countries along the South and Central American Pacific shoreline averaging 4,204,000 barrels of oil or 1,261,200,000 tons and 96,294 Mtoe (million tons of oil equivalent) per year. This is a total of 1,357,494,000 tons of oil and natural gas taken out of the ground every year. This is an enormous amount of protection from the mantle, releasing heat that would not be released if we did not interfere with the natural process. These numbers do not even include the major oil-producing areas that surround the north and eastern parts of the Ring of Fire, namely Russia and the United States.

It is very hard, if not impossible, for the human mind to comprehend large quantities or distances. To comprehend billions of barrels of oil and trillions of cubic feet of natural gas is an impossible task. However, this is something that we must do somehow. Swimming pools were used as an example earlier, but we really can't comprehend its volume or a large number of swimming pools used as an example of oil being removed from any given location. Can we comprehend the vast national debt that is climbing beyond $18 trillion? We talk about stars being light-years away, not realizing what a light-year is, or for that matter the speed that light travels in a second. How then can we visualize a pipeline that crosses over eight hundred miles between Prudhoe Bay, Alaska in the north to Valdez, Alaska in the south with hundreds of feeder pipes filled with oil?

How about visualizing the Earth as in the movie *The Return of the Jedi*, which shows Death Star II in *Episode VI* outside the Forest Moon of Endor that is seen unfinished. Our planet is becoming like this with the removal of vast quantities of oil and natural gas being removed all over the world. We are also removing huge amounts of iron ore, phosphorus, copper, and many other elements. I could go on and on with this. Be honest about our planet that this is what is happening. Realize it before we get to the point of no return—when people must suffer its tremendous forces that it will release.

Before I started this research on global warming many years ago, realizing that the heat from the core could be causing so much heating of the Earth from the drilling of oil and natural gas wells, I never thought that the thousands of miles of pipelines are radiating so much heat into our environment. I was always for the Keystone Pipeline, but now I can see how the heat from these pipes can cause irreversible damage over time to our environment. I know this pipeline is only one, but tens of thousands of miles worldwide can and will do a great deal of harm. It must stop sometime before we reach the point of no return. This radiating heat from this pipeline could cause substantial problems to our food supply. So could the so-called heat islands of cities and millions of miles of roadways. However, I am more concerned with the removal of protection of the heat from below than from above.

Let's review a few earthquakes that have happened. On New Year's Eve in Ohio, a 4.0 magnitude earthquake occurred. According to scientists, it was not natural. It was at a depth of two miles. Won-Young Kim, a professor of seismology at Columbia University, stated that earthquakes do not occur naturally at that depth. It is believed the earthquake was caused by high-pressure liquid injected into oil and natural gas wells. I believe the problem has occurred underground before, and this high-pressure injection of liquid (who knows what it is composed of) just aggravated the situation more.

Near Oklahoma City in 2014, at least four earthquakes from 2.9 to 4.3 magnitudes struck about thirty miles north. Why? Oklahoma did not have intense earthquakes until recently. Up until around 2009, Oklahoma had an average of only two earthquakes per year that was higher than a magnitude of 3.0. However, in the past six years they have had thousands of earthquakes, and this has made it the second most active state in the continental United States just behind California. This area of the United States was stable and void of any major earthquakes. I believe this is a direct result from fracking. I will deal with fracking later on in this book, but at this time, I would like to remind people that *fracking* means "to break into pieces." The very word gives away what is happening below the surface.

There are increasing numbers of earthquakes in states that once were considered stable regions, but these states in recent years have experienced more and more tremors, namely Alabama, Arkansas, Colorado, Ohio, Kansas, New Mexico, Oklahoma, and Texas.

I am going to conclude this section with an example of the wooden block game of Jenga, a game that involves wooden blocks stacked one on top of another to build a tower. Then each player removes a block until a player pulls out a block causing the tower to collapse. If we could visually replace the blocks with blocks containing oil, it could represent what we are doing to the Earth. This leaves us with the crucial question: How many blocks of oil can be removed until it all comes down?

Dried Out

Here are some examples of historical events that have occurred that can be contributed to the removal of oil and natural gas.

The solving of the Dust Bowl in the 1930s was only a Band-Aid as to what really happened. It was effective, but it only covered up the problem that was triggering the drying out of the land in the first place. As with many things, people tend to solve problems only if they can see it, hear it, or feel it. There is another way that must be considered. For example, very few people have studied or understand quantum mechanics, yet you can determine the results even when you cannot see the actual process. Sometimes this process is called *cause and effect*. Just because you can see results doesn't always mean the underlying problem has been eradicated. The Dust Bowl can be put in this category. The Dust Bowl started sometime in 1930 and continued until it ended in 1939. The factors that contributed to it started several years earlier.

The first thing that can be contributed to the Dust Bowl was the beginning of oil drilling in Oklahoma and Texas. Oklahoma's first commercial oil well was in 1897, and within ten years, the state became the largest producer of oil in the world. Texas's main drilling followed around 1899–1901. This was the beginning of the release of heat from beneath the crust of the Earth that did not naturally occur. It worked like this: when oil is first extracted, heat is brought to the surface locally through the heat in the oil and natural gas. Natural gas was a byproduct at this time and considered more of a nuisance, as it is in many parts of the world today (Fig. 12). Gas is still burned as a byproduct, which can be seen happening all over the world. This in itself creates heat above the ground as do wide fires, burning buildings, furnaces, and volcanoes. When this protection or insulation of oil and natural gas (gas absorbs many times more heat than oil) is removed, the rocks then absorb this heat and release it above ground. Rocks are not even close to protecting us from the heat rising from the interior of the Earth as oil, and especially natural gas. When oil and natural gas are removed, the heat is absorbed through the crust for a period of time. How much and how long is dependent on the depth of the well being drilled. Shallow wells release heat much quicker than deeper wells. Another factor is the quantity being removed. Also, a determining factor is the amount of time spent in extracting oil and natural gas. After this process runs its cycle and heat will no longer be released to the surface as it once was, the ground will be totally dependent on the seasonal heat from the Sun.

Figure No. 12: Natural gas burning from oil well.

The next factor, which created the Dust Bowl, is the extreme amount of oil extracted in the following years after it was first discovered in this area. It became highly profitable to extract oil out of this area because of the invention of the automobile and the increase in the use of oil needed in World War I. All this dried out topsoil, lead to dust storms for over a decade. It took approximately thirty years for the heat to rise and dry out the topsoil.

Keep in mind that these are very subtle changes taking many years for the heat to rise from below the surface. A volcano releases its heat in a short period of time. It seems to take the removal of oil and natural gas around thirty years for this unprotected heat to rise to the surface to affect the climate. This time period could easily be shortened depending on the depth of oil being extracted and the amount. Case in point is hydraulic fracturing, which I will cover later in this book.

In the last few years, California has experienced extreme drought. There have been many heat advisories in Los Angeles in recent years. California has been pumping oil and natural gas out of ground and offshore for over a century. In the early discovery of oil in 1850s and 1860s, oil was noticed to be bubbling up from the surface. This oil that was found near the surface was sticky and would not flow through pipelines until it was heated especially in the winter time. Much of this oil was used as asphalt, also known as *bitumen*, to pave roads. A major discovery of oil deposits was found in 1890 near Los Angeles, which was later called the Los Angeles City Oil Field in 1901. At that time, it became the largest producer of oil and natural gas in California. During this boom, there were two hundred oil companies extracting oil and natural gas from this area. Now there is only one. This is very significant in relating to my new approach toward global warming. The enormous quantities of oil and natural gas extracted out of this area have depleted the region of its protection from the extreme heat source just under their feet. California in 1900 produced four million barrels of oil. Keep in mind that this does not include millions of cubic feet of natural gas escaping into the atmosphere as vented natural gas, which was not used, and there were probably another million barrels wasted due to poor drilling methods.

California and Oklahoma started trading places for the number one oil-producing state until 1930 when Oklahoma experienced the Dust Bowl (Fig. 13). California was producing thirty-four million barrels of oil by 1904, and by 1910, it was producing seventy-eight million barrels per year. Oil was found in the San Joaquin Valley in 1899, and by 1903, oil was discovered on the west bank of the Kern River just north of Bakersfield. After that, 70 percent of California's oil was taken from these wells. There are pictures of spectacular gushers taken from all over the valley demonstrating the tremendous loss and environmental destruction that oil and natural gas have been causing to our planet for over 150 years, and that was before it was transported. Today the San Joaquin Valley oil field is producing over 515,000 barrels of oil per day or approximately 188 million barrels per year. That is over one billion cubic feet of displacement under the Earth per year. Add that to the 248 billion cubic feet of natural gas extracted every year, and you have a tremendous amount of protection destroyed. Keep in mind that this removal of oil and natural gas cannot be replaced. California as well as other regions all over the world that are following the same course must adapt to a new, more stressful way of life. This is oil taken out of the San Joaquin Valley alone and does not take in the rest of California.

Figure No. 13: Oil derricks Huntington Beach, California

California has put in restrictions on new oil and natural gas ever since the Santa Barbara oil spill in 1969, which released an estimated almost one hundred thousand barrels of oil into the Santa Barbara Channel and became the largest oil spill in the United States, only to be over shadowed by the

1989 Exxon Valdes and the 2010 Deepwater Horizon spills. Since the Santa Barbara oil spill, California has passed several bans on offshore drilling like the 1994 California Coastal Sanctuary Act, which prohibited new leasing of state offshore tracts that can stretch three miles offshore.

Then President George H. W. Bush issued an executive moratorium banning new federal leasing through the year 2000 not only in California but also in other offshore areas. These areas were extended to 2012 by President Bill Clinton but later rescinded by President George W. Bush in 2008. I mention this because we become satisfied when the government or entities restrict things but do not remove things. I say this because oil and natural gas companies are still continuing to do the same thing even when restricted. They do this by horizontal drilling outside their lease, which could also be by oil simply draining into existing wells from outside their lease area. The second way these new restrictions are ineffective is that there is enough oil and natural gas in a restricted lease to last a very long time, so in order to satisfy their opponents, they decided to approach these limits at a later date with more lenient government officials. As you can see, this has already happened in California.

There is a huge amount of protection being removed between the Earth's surface and the heat from the mantle. Is it any wonder that California is experiencing heat advisories and drought? We can easily compare this same scenario to other regions in the world.

Has anyone noticed that the regions of the world that do not produce oil and natural gas are not experiencing the same extreme weather conditions as those that do? As the consumption of oil and natural gas increases worldwide, it will only increase the warming of our globe.

Other historical catastrophic events that can be traced back to the extraction of oil and natural gas are hurricanes, tornadoes, earthquakes, and even El Niño and La Niña.

Melt It Down

The glaciers are melting! The glaciers are melting! It sounds like Paul Revere. As it was true that the British were coming, it is also true that the glaciers are melting, at least for now.

Okay, let us run this scenario. You go to Alaska, drill a new well for oil and natural gas, and then begin pumping oil out of the ground for sixty years. At first, you pump nine hundred barrels a day, equal to approximately thirty-seven thousand gallons or about 5,050 cubic feet. Using the swimming pool to relate, figuring approximately twenty-thousand gallons per pool, this would be approximately two pools per day back in 1957. Today the Trans-Atlantic pipeline carries 850,000 barrels of oil a day (1,827 swimming pools per day)—or put in perspective: 1827 swimming pools × 365 days per year × 60 years. You get around forty million swimming pools of oil out of Alaska alone, not to mention the rest of the world. This does not consider the volume of natural gas being removed, adding trillions of cubic feet.

Figure No. 14: Melting Ice Cubes

Let us see if we can explain glacier melting my way. A glacier is no more than a giant ice cube (Fig. 14). An ice cube melts from the bottom up not the top down, as some would assume from just observing it. It can be noticed in aerial pictures after snow storms that the snow melts on cold surfaces first like road and bridges; however, it forms ice in its place. This is due to the cold surface it rests on, sometimes called *black ice*, especially on bridges because the underside of bridges are open and remains cold. Homes have similar areas under the foundations and around openings that allow cold air to penetrate inside called *void space*. The ground radiates heat as do roofs of homes and businesses and melts the ice. At the North Pole, the average temperature during the summer months of June, July, and August is 0 degrees Celsius (32 degrees Fahrenheit), hardly enough to melt ice from the surface. If you remove the protection from the core and allow heat to radiate constantly toward the surface 24-7, it would cause the ice to melt. Many scientists say the atmosphere is heating up due to climate causing the ice to melt. An average temperature during the warmest part of the year is hardly a reason to prove global warming.

Scientists also say the oceans are warming, causing the ice in the Arctic to melt. They have documented the retreat of ice for years, but most of these are on rock. Take for instance Austfonna, which is an ice cap located on Nordaustlandet in the Svalbard archipelago in Norway. It covers an area of 3,129 square miles (8,105 square kilometers), which includes Vegafonna, a huge glacier that rests on a frozen rock north of the Arctic circle in Norway's Svalbard chain of islands. NASA has documented the melting of this polar ice by satellite for many years, but the reason is under the ice, not above.

This is a very complicated system. The atmosphere does come into effect, but in a way, that is not being accepted today. Greenhouse gases are not contributing to the melting of snow and ice around the North Pole at this time. It will not come into effect for another fifty to one hundred years, depending on how we change our way of doing things. What should be of immediate concern is the removal of vast quantities of oil and natural gas from these regions.

You probably have caught the statement that the glacier referred to in Norway is on frozen rock. The question is how the glacier can melt from the bottom up if it is on a frozen rock. My theory does not contradict itself. Here is my take on this solution.

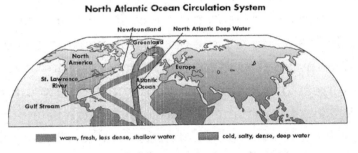

Figure No. 15: Thermohaline Circulation

There is a climactic system called the *thermohaline circulation* (Fig. 15). This system is naturally driven by the Sun heating up shallow currents in the tropical oceans. It is being increased by the removal of oil and natural gas in these regions, like India, China, and Australia, just to mention a few that have increased their production of oil and natural gas substantially in the last few years. Heat is absorbed in the tropics and carried by ocean currents from the warm waters of the tropics, namely the Gulf Stream that flows into the North Atlantic. This heat is then vented into the atmosphere. As these warm waters flows into the North Atlantic, it transfers its heat into the atmosphere and becomes cooler. The water becomes saltier and sinks to the depths of the ocean before upwelling back to the surface. Please note that when frozen, water releases salt; and when it melts, it is free of salt. This reaction causes the heavier salt water to sink and the less saline water to rise. According to NOAA, this process is dynamic and has been known to dramatically shift, as it appears to have done in the past. They say it can cause major changes in climate over a very short period of time (ten to twenty years). Therefore, what has changed is the removal of the insulation and protection from the extreme heat of the core that oil and natural gas is preventing. This protective insulation is not only being removed in the North Atlantic but also in the Gulf of Alaska. This is upsetting the natural circulation of the ocean currents. The heat is being extended farther north than normal and is being heated more than it normally would be. This warmer air off the oceans causes the snow and ice to melt. You can perform your own experiment by taking a hair dryer at a safe, comfortable distance. However, moving it closer without increasing its temperature will feel much hotter as if you would have increased the temperature. Therefore, as the currents get warmer, they will extend further and further in the arctic regions, thus melting more snow and ice.

A word of caution: do not be too sure this will continue indefinitely. I will explain this later. The average temperature at the North Pole is 32 degrees

Fahrenheit (0 degrees Celsius). During summer (June, July, and August), the highest temperature recorded is 41 degrees Fahrenheit or 5 degrees Celsius. Keep in mind that this is an average; it is not always this temperature. This is hardly enough to justify the melting of eons of ice glaciers.

Scientists believe this warming is causing the ice to melt because the last four years was the warmest on record since record keeping started back in 1895, and it was the thirty-sixth consecutive year of an average world temperature of 57 degrees Fahrenheit, this according to a NOAA report. I do not question the report, only the method scientists use to determine how it is being applied.

Therefore, I ask this question: How could temperatures that last only a short time during the summer months (remember these are average temperatures) melt huge glaciers of ice so quickly? One must take many factors into consideration to make this occur—not just temperatures but pressure, gravity, wind, earthquakes (which I will tackle later), and of course, the heat from the core. If you set a block of ice outside during freezing temperatures and it warms up to just above freezing, it will melt slightly; but if the temperature goes back to freezing, the melted water will refreeze. If the block were slightly heated from underneath while you still have freezing temperatures above, it would slowly melt.

I stated earlier that this melting would not continue indefinitely. I will explain it now. When heat is released through the extraction of oil and natural gas, it is lost and cannot be returned. It is estimated that four and a half million years ago, the crust of the Earth was about four and a half miles thick, and it would have been very warm on the surface. Humidity would have been very high to the point where rain would not settle on the surface, if it rained at all. Over the years, heat escaped to the surface, and the rocks cooled. Today the crust is estimated to be around twenty-five miles thick. As the thickness of the crust increases, there will be less heat able to rise to the surface; hence, the surface will be colder (remember the void space under structures and under bridges?). It will take time for this to become a serious reality. It could be ten, fifteen, twenty, or one hundred years. At the present time, it is the heat that should be of great concern.

Scientists have determined that methane is bubbling out of the Arctic at a much faster rate than previously thought. Methane is almost thirty times more effective at absorbing heat than carbon dioxide. It is just this conclusion that the heat-absorbing characteristics of methane are not only being removed but also being released from the ground, eliminating the insulating qualities of it. However, one might say that if it has such an effective trapping ability of heat, it could then be balanced out. Heat can be absorbed into the atmosphere, but from the core, it cannot be replaced. This is another reason this could bring on another ice age much sooner than would naturally occur. Again, this is a long way off to be of concern for the immediate future.

Heat escaping from the core should be our immediate concern. That is what is causing the ice to melt. This melting is even a more crucial concern among animals that need cold to survive. Animals such as seals, moose, polar bears, musk oxen (Fig.16), and penguins, just to name a few, are being invaded with parasites, ticks, and lungworms that normally would not survive in colder climates. These diseases are taking a tremendous toll on wildlife all over the world that otherwise would not be an issue.

Figure No. 16: Musk Oxen

The enormous amount of oil and natural gas being extracted out of Siberia, Alaska, and the North Sea is causing much of the rise in temperature. Most scientists are using the greenhouse effect to explain the increase, but if that were so, then places that have large volumes of pollution expelling from factories and cars would be extremely hot. They would not have the season of a long cold climate to cool things down as it does in the northern latitudes, which would cause much higher temperatures than they presently have.

Another problem we face may not be readily seen. When snow and ice melt, this gives oil and natural gas corporations easier access to land that they can drill, thereby extracting more insulation out of the Earth and making it much hotter. It becomes very difficult for these companies not to drill for more and more oil due to the fact that the northern latitudes is believed to store up to 30 percent of the Earth's oil and natural gas reserves. This is the reason why the Earth is so successful in holding on to its heat reserve. Other planets in our solar system are not as efficient in this area. They do not have the insulating ability of Earth. It seems that in areas where insulation from the core is needed, it is provided. In cold climates, heat would quickly be absorbed and be forever lost if it were not for the vast amounts of oil and natural gas below the surface protecting us. On the other hand, warmer climates such as Saudi Arabia, Kuwait, etc. would literally be too hot for humans to exist in everyday life. Recently, Iraq closed businesses and schools due to the heat, which reached temperatures as high as 126 degrees Fahrenheit (52 degrees Celsius). This could have been, according to my theory, the direct result of extracting a greater amount of oil and natural gas after the Gulf War. The quicker oil and natural gas is removed from the ground; the more heat is released in any given area.

If the Earth's atmosphere is composed of 78 percent nitrogen, 21 percent oxygen, 0.94 percent argon, only 0.03 percent carbon dioxide, and finally 0.03 percent of other gases, how could it possibly be imagined that the small amount of carbon dioxide could have such an effect on our climate? Oil and natural gas, however, are being removed all over the world in large quantities on land and at sea. In some cases, they are removing oil and natural gas over very large tracts of land in California, Texas, Gulf of Mexico, Siberia, North Sea, Saudi Arabia, and the Pacific Ocean.

One can easily show that there is a rise in temperature and the melting of snow and ice and even the melting of glaciers right before our eyes. But why? What is causing it? I contend it is the removal of the insulating power of oil and especially natural gas. Global warming is just that: global warming, or the heating up of the Earth by the core.

Go Direct

The most damaging process in global warming that is being developed today, besides the volume of oil and natural gas being removed from the Earth, is geothermal energy (Fig. 17). The process involves drilling into the Earth to extract heat from hot rocks close to the mantle. This tapping into the Earth is releasing enormous quantities of heat to the surface to run steam generators in order to supply businesses and homes with energy. It is this process that is causing the crust of the Earth to become thicker by eliminating heat from the mantle over time. When rocks that are composed of different types of minerals are heated, they expand; and when cooled, they contract, which could cause tremors or even earthquakes. The removal of oil and natural gas also releases heat that protects us from the Earth's core, which could cause tremors or even earthquakes. The released heat in a single well is unnoticeable in most, if not all, cases; only instruments can detect this heat. However, the small volume of heat in each well is the culprit in releasing heat while extracting oil and natural gas, whereas geothermal energy is massive and releases heat very quickly.

Figure No. 17: Geothermal Energy Plant

Exploratory wells have determined that this would not be economically feasible to pursue, except in places that are very close to the hot lava of the mantle. It has already occurred in Germany where an experimental geothermal plant was turned on and caused tremors approximately one mile away. These tremors are thought to be the result of the construction of a geothermal plant. Also in Basel, Switzerland, another plant was turned on and also caused tremors within a mile of the plant.

The idea was to drill into a hot layer of rock located three miles below the surface made of water-free, impermeable rock. Water is injected down into the borehole in order to generate power. When the water reaches the impermeable rock, it turns into steam. The expansion of this creates enormous pressure, forcing the steam back up to the surface; a visual example is seen in the Old Faithful geyser in Yellowstone National Park located in Wyoming. This steam then would be used to spin electrical generators. Any leftover hot water would be used to heat homes and businesses. These plants remove tremendous amounts of heat from the dry rock beneath the surface down as deep as six miles.

These geothermal plants accelerate the flow of heat much quicker than oil and natural gas. Also, smaller systems are being used throughout the world for individual homes and businesses. They are simple in nature, using an enclosed loop of one-inch polyethylene pipes that are filled with water. Heat is still being removed from the ground, which gets it heat from the mantle in the Earth. If this system became commonplace and used in millions of homes and businesses all over the world, it would further deplete the heat needed to ensure a normal process of heat rising up from the mantle and core of the Earth.

This heat removal also causes an increase in the shrinking of the mantle. This shrinking of the mantle then causes a larger gap between the crust of the Earth and the core and in turn causes the settlement to occur, which causes instability and earthquakes.

That is a lot to take in, but global warming is much like a chain reaction.

The larger question that may be difficult to determine is whether or not this decrease in the size of the mantle is local or is consistent throughout the entire Earth. If the decrease is consistent throughout the entire Earth, then it would seem—if put in proper perspective with its size—that such extraction of heat would probably be negligible over a period of time. However, if this shrinking is local or sporadic around the mantle, then there is a much greater likelihood that the settlement of the crust would fall into the mantle and cause tremors and earthquakes. This action would cause a greater reaction in a future earthquake because the core would be uneven or hilly, which would result in the settling of the Earth's crust toward the center at a greater height. This seems to be the case with the geothermal plants in Germany and Switzerland. When they drilled these boreholes, they struck extreme heat quicker in some areas than others, which leads one to believe that the mantle and core are not consistent throughout.

One might speculate that there are plants elsewhere that have not caused noticeable earthquakes; one such plant in the United States does not appear to have caused any adverse tremors. This may be because the plant is built on a thicker, more substantial part of the Earth's crust. However, this cannot be construed as not having a possible effect in the future or that the removal of heat directly from the interior of the Earth may be causing irreparable damage. The crux of geothermal energy is that it cannot be corrected or renewed, for once the heat is removed out of the Earth, it can never be replaced. This leads us to another problem that will be presented in another chapter that is diametrically opposite to that which has been previously presented here.

I cannot leave this subject without mentioning Greenland's and Iceland's use of geothermal energy. Greenland is located between latitude 59 degrees and 82 degrees north and longitude 11 degrees and 74 degrees west. It is the largest island in the world. A layer of ice covers the interior of the island. The coast is free of ice. The extreme north of Greenland is called Perry Land and is not covered with ice. They say it is because of its dry weather. Why? I contend it is because of heat rising from the mantle. This is why they have been so successful in tapping into this heat source because it is relatively closer to the surface than some other areas in the world. One such event occurred in Iceland in which a well was drilled near a volcano (brilliant, huh?). They then had to stop drilling because less than half way down from their projected depth of fifteen thousand feet, red hot magma entered into the well, forcing the team to stop drilling. This heat from the magma would have produced a tremendous potential to run turbines to generate electricity. However, it does not take into the fact that once again, the extraction of heat cannot be replaced and will eventually cause irreparable damage to their climate in future years sooner not later.

Iceland is located close to the Arctic Circle and is one of most volcanic areas in the world. It is unusually warm to be so far north. They say it is warmer because of the Gulf Stream, which keeps its climate milder than otherwise would be for an area so far north. More will be said about the Gulf Stream later.

These two countries are doing the world a future favor, but in the process, they are destroying their climate. Here is my take on this. The geothermal plants and hydrocarbon plants are sucking heat from below the surface at an alarming rate. This heat has kept Iceland and the coast of Greenland's climate milder for centuries. Eventually, this will stop sending heat up from these hot areas. When this happens, the soil will not thaw enough to allow the snow to melt, and then it will turn into glacier-like conditions and will change this milder climate forever. As a result, in favor of the world's glaciers, they will increase and not melt. This is happening before our eyes now. Scientists have been monitoring an unusually cold part of the water in the North Atlantic south of Greenland. They call this a "blob." As I mentioned earlier in the chapter on glaciers that referred to a climatic system called the thermohaline circulation, the waters of the North Atlantic was being heated because of the removal of oil and natural gas by drilling and has extended the warm water further north than usual. As these warmer waters extend further north, it will also cause the return of the currents back south to be colder due to a lower temperature in higher latitudes, thereby making the water south of Greenland colder. Eventually, the impact of water circulating back to the tropics will reduce the amount of heat being naturally produced in these warmer climates. The cooling of these waters will have the opposite effect of what we are experiencing now. That is, it could make areas in the north no longer have their mild climate.

The real danger for humanity is removing the insulation that protects us from the extreme heat of the core. The countries that are above the 35 degrees parallel (the use of 35 degrees parallel is a reference point, not a fact) are in danger of losing the natural process of heat from the mantle and becoming much colder, bringing on the beginning of a new ice age. Countries like Canada, Great Britain, Northern Europe, and Russia are the most vulnerable if the continuation of this removal of heat is not stopped.

The extraction of the insulating properties of the oil and natural gas protecting these countries will at first cause them to warm, such as the melting of ice. After that, they will begin to experience colder winters and shorter summers. The soil will not thaw soon enough to allow for efficient planting of fields to harvest food.

Greenland and Iceland will be the earliest signs of major irreversible climate change. This could be the beginning of another ice age. Anything above 35 degrees parallel will have colder and longer winters, which will not allow the ground to thaw soon enough, resulting in more snow and ice each year. It cannot be stressed enough the danger that geothermal energy is doing to the Earth.

In 2009, the borehole at Krafla in Iceland tapped into the Earth's magma at only 6,890 feet (2,100 meters) deep. A very rare occurrence at this time, but if it proves that the investment is profitable, it will only be the first of many.

A study was put together by the MITRE Corporation for the US Department of Energy (DOE) through the Geothermal Technologies Office (GTO) the Office of Energy Efficiency and Renewable Energy. They were to study the developing methods for mining Earth's heat to depths of 0.6 to 3.1 miles (1 to 5 kilometers) in hot, dry rock. The conclusion was that it would be too expensive to justify the cost of the small amount of heat it would recover. Their formulas and examples used in their study could prove how much heat is escaping from any given well.

The study also mentions different types of drilling techniques to minimize rock damage. I will approach this when discussing earthquakes. The study mentions several factors that need to be examined further. For instance, the extractable heat increases linearly with depth, which alerts my theory that when drilling deeper and deeper holes for oil and natural gas, the heat factor will increase substantially. The hot crust is buoyant so that it rises and is eroded at rates $\approx$).1-1 km/Myr (it also moves laterally at rates of $\approx$10 km/Myr due to plate tectonics). I mention this from the study to ask the question: Are we shortening the length time it takes to lose natural heat? If so, we could be bringing on the coming ice age by millions of years while at the same time having to experience unbearable hot conditions all over the globe for centuries, even to the point of creating inhabitable zones around the world.

Heat-Seeking Missiles

Hurricanes are like heat-seeking missiles; they have an insatiable appetite for heat. Without heat, they cannot exist. Without heat, a hurricane can't even come into existence. Heat is the fuel that energizes a hurricane.

As of the writing of this book, a hurricane called Patricia developed very quickly, actually surprising meteorologists off the coast of Mexico. It was a Category 5 hurricane, the highest designation on the Saffir-Simpson Scale. This scale is used to measure hurricane wind strength, with record-breaking winds of 200 mph (325 kph). This was the strongest storm ever recorded in history either in the Pacific Ocean or the Atlantic Ocean. Fortunately, the storm weakened before it came ashore, but it still had a speed of 165 mph (266 kph). It also did not follow its projected path toward the more densely populated area of the port city of Manzanilla.

I covered why this hurricane could have developed earlier in this book, but I want to take this further and notice that when this storm crossed Mexico, it did not head for the warm waters of the Gulf of Mexico. It turned northeast toward Texas. The question of why this storm would deviate from the more logical track can be determined in the quantities of oil and natural gas production between Mexico and Texas.

Mexico produces an average of three billion barrels of oil per year, while Texas produces more than 351 billion barrels per year and over six trillion cubic feet of natural gas. That is an awful lot of oil and natural gas being removed, thereby releasing more heat from the interior of the Earth's surface.

I realize air currents and pressure cause a lot of the direction that hurricanes follow; but they also are caused by heat, which is generated mostly from a source that is only twenty-three miles below our feet. The science of weather needs to encompass this process in their predictions. Many examples can be used to show how the removal over time has created a massive hurricane that typically would not have happened if oil and natural gas were not removed.

Other hurricanes can be traced to the removal of oil and natural gas from the Earth that is leaving us unprotected from its heat. Here are some examples that I will use, which are historical in nature. One of the earliest hurricanes in the past one hundred years, which was not very well documented, is called Labor Day Hurricane of 1935. It was the first Category 5 to be recorded to make landfall in the United States. Why? Let's look at it with a different approach. El Niño's had a longer existence than usual before and after 1935. Normally, El Niño events generally are shorter than La Niña; but in this case, they were two of the longest active El Niño events that occurred between 1918–1949. The particular years were between 1929–1932 and 1939–1942. In the middle (1935) was the 1935 Labor Day hurricane. This means that the Pacific was much warmer (El Niño) over a longer period of time. In 1935, it was warmer in the Atlantic Basin, hence La Niña. The reason can be traced to the production of oil and natural gas in the Middle East. Here is why.

The United States was not the only new producer of oil and natural gas during this time. England and France were very much in earnest in finding oil for the new technology that helped win World War I, namely the internal combustion engine. The oil they found was in the Middle East in 1927. Then soon after, two pipelines were constructed, one by the French through Syria and Lebanon ending at Tripoli on the Lebanese coast. The second pipeline was built by the British and Iraqis terminating at Haifa. Each pipeline carried 13.7 million barrels of oil or 27.4 million barrels combined a year. This may not seem like a lot according to today's production, but there is a huge variable in this. When oil was found, there were enormous gushers of oil released into the area; just for the record, this has occurred throughout the search for oil worldwide (Fig. 18). These gushers sometimes took weeks to bring them under control while releasing large amounts of hot oil and natural gas. I believe this caused an unusually warming trend across Africa into the Atlantic Basin, which in turn produced the strongest and most intense hurricane to come ashore in the United States.

Figure No. 18: Oil gusher at Signal Hill, California near Long Beach. One of the most productive in the world. 1920s

Hurricane Camille was another Category 5 made landfall near the mouth of the Mississippi River on August 17, 1969. It also quickly strengthened over areas in the Gulf of Mexico known for significant amounts of oil and natural gas wells. Hurricane Camille originated off the western coast of Africa very similar to the 1935 hurricane with a warming trend across Africa into the Atlantic basin.

Hurricane Andrew, a Category 5, went across the southern tip of Florida in August 1992. Why did it enter here over Florida? What has changed over time to cause this catastrophic storm to go over an area that is not known for such an intense storm? This hurricane happened during what is sometimes called a "cool phase," a relative calm period. If this was a cool period and hurricanes are notoriously noted to seek heat, where did the heat come from? It can be tracked back to oil drilling.

Two oil wells in Southern Florida's Collier County began producing over 3.4 million barrels of oil during 1984 to 1992, and this was only two wells. Oil drilling in Florida was encouraged by the governor and cabinet of Florida more than seventy years ago, at which time they offered fifty thousand dollars to the first person or company to find oil in Florida. In 1943 Humble Oil and Refining Co., later to become Exxon, struck oil near Immokolee, thereby winning the fifty thousand dollars. They drilled many wells in a narrow swatch of land called the Sunniland Trend that runs diagonally from Fort Myers to Miami. This geological feature is some 150 miles long and twenty miles wide. It contains oil and natural gas deposits about two miles underground. Florida's soil is very porous, which would allow heat to rise up faster than most areas being drilled in the world. It was this extra heat injection between 1984-1992 that bloated Hurricane Andrew into to a full blown Category 5. Keep in mind that it is not the heat from the oil and natural gas that is the problem; it is the removal of the insulating properties of these materials that provide us with the needed protection from the extreme heat under our feet.

It is my hope that Florida politicians analyze this and refrain from removing oil and natural gas from Florida soil and off its coast in the future. We should be protecting the ideal climate that Florida is so well noted for from the extreme heat that lies so close under our feet. I must say that the Sun causes enough heat without making the matter worse when we have the power to stop it from happening in the first place.

Hurricane Gilbert (1988) and Hurricane Dean (2007), both a Category 5, had similar tracks that brought them into the Caribbean Sea and the Gulf of Mexico. Again, I ask why.

Mexico found significant quantities of oil and natural gas in the Gulf of Mexico in 1976. From 1979 until 2006, Mexico has extracted over eleven billion barrels of oil and 4.5 billion cubic feet of natural gas. This oil drilling complex, named Cantaroll Field, in the Gulf of Mexico is a city in itself with all the services available as if it were land-based. This oil field had 190 oil wells. In 2003, its production was the quickest producer of oil and natural gas in the world, eclipsed only by the Ghawar Field in Saudi Arabia. So one can see that the removal of oil and natural gas has a significant effect on the unnatural intensity of these hurricanes. Production started in 1979, and nine years later, Mexico endures a Category 5 hurricane, Gilbert. Then in 2003, these oil fields became one of the fastest producers of oil and natural gas in the world. Four years later, Mexico gets another Category 5 hurricane called Dean. Is this a coincidence? I think not.

In 2005, Hurricane Katrina became a Category 5 on August 28; its center was located 195 miles southeast of the mouth of the Mississippi River in the Gulf of Mexico. It come ashore as a Category 3 (125 mph) and is considered one of the most devastating hurricanes in the history of the United States, but that is relative to the area it comes ashore and how prepared the people are. I must say that in the case of Katrina, the dikes holding back the force of water were not very well prepared. Here again, the increase in intensity was due to the thousands of wells pulling oil and natural gas from under the Gulf of Mexico releasing heat that would otherwise be absorbing heat.

I cannot leave this section unless I mention Hurricane Sandy. It comes ashore on October 29, 2012, near Brigantine, New Jersey, just north of Atlantic City. Though it did not come ashore as a Category 5, its uniqueness was when it came ashore. Hurricane Sandy—instead of continuing northeast and expiring out into the ocean, which is natural for a hurricane to do—instead took a sharp left turn toward the west-northwest. Here again, why? Some would say it was because of the heat generated by the large metropolitan area of New Jersey and New York, but this storm went further inland and caused much damage in twenty-four states all the way across the Appalachia Mountains as far as Michigan and Wisconsin before its demise. I believe it was because of the extraction of oil and natural gas due to the increased usage of fracking that has made the United States the largest producer of oil and natural gas in the world. The removal of the protection of insulating qualities of oil and natural gas from the hot mantle are making the interior States a "hot zone."

There are several things all these hurricanes have in common. They quickly strengthened in intensity. They occurred soon after the beginning of significant extraction of oil and natural gas. They all deviated from the typical development of a hurricane and direction. During the last decade (2000–2009), we had more Category 5 hurricanes than any decade in recorded history with eight. These were Isabel (2003), Ivan (2004), Emily (2005), Katrina (2005), Rita (2005), Wilma (2005), Dean (2007), and Felix (2007). The previous decades with most Category 5 hurricanes were in 1930–1939 and 1960–1969. They seem to have one thing in common. They all came into existence after substantial increases in oil and natural gas production.

One thing is for certain: when heat rises from the surface, things change. And most of the time, that change is not good.

The Tops Keep Spinning

In Oklahoma, Nebraska, Kansas, Texas, and throughout the Midwest, the increase in intensity and quantity of tornados is increasing at a rate never before recorded in history. Eventually, there may not be a place safe enough to exist (Fig. 19a). Like hurricanes, the increase in these sweepers of destruction can be linked with the removal of massive amounts of oil and natural gas.

Figure No. 19a: Super-cell Thunderstorm

Just recently, a tornado occurred in Denair, California. The news article started like this: "DENAIR, Calif.

(AP)–A rare tornado struck a Central California town, tearing roofing and walls, knocking down trees and power lines and damaging gas lines."

The key word here is "rare." I could list many other tornado events throughout the world that use this word to describe them, plus the most popular word "historical"; but normally, they use these words to describe the amount of damage they have done. The one statistic that stands out is that we are having as many as eight hundred tornadoes per year compared to the norm of a hundred on average (Fig. 19b). These are relatively weak tornados as measured on the Enhanced Fujita scale, a scale used to classify the intensity of a tornado, with winds between sixty-five and eighty-five miles per hour, labeled as an EF0. Of course, there has also been an increase in stronger tornadoes. I am concerned with the 800 percent increase in recent time.

Figure No. 19b: Multiple tornadoes

The Sun has not changed its location; neither has it increased its heat. The Sun still sends light at the same time and at the same distance from the Earth that it has for thousands of years. So what has changed? You guessed it: the removal of oil and natural gas protecting us from the heat of from the core. It is this subtle release of heat from thousands of oil wells over large areas throughout the Midwest.

Researchers are reconsidering their climate computer models because they are confused with the environmental conditions that are not showing these unusual events. They are still examining changes in large-scale atmospheric conditions relevant to major destructive tornadoes. These conditions evolve wind speed change and wind shear and air turbulence. All these condition are occurring above the surface of the Earth. No one is looking at what is happening below the surface. Add the variable of localized temperature and they will see that the major cause of this increase in the quantity of weak EF0 tornadoes are directly a result of an increase where well drilling is occurring throughout any given area.

Most tornadoes occur in April, May, and June. The strongest develop in April and the most in May. These are hardly the hottest months in the Midwest, but tornadoes still need heat. The ground is warmer during the winter months due to the extra heat escaping from thousands of oil wells. I am not suggesting a person can physically fell the change, but climate conditions over large areas are sensitive to small increases of heat.

In a document put out by the NOAA National Severe Storms Laboratory, it asked the question "How do tornados form?" and the answer they give is "The truth is we don't fully understand."

They photograph tornadoes. They can determine their wind speed. They identify tornadoes with names like "beaver's tail," "wall cloud," "rear flank down-draft," "condensation funnel," and "inflow bands," spending millions of dollars and man-hours and still cannot find the real reason why.

Recently, NOAA completed a study from May 10, 2009 to June13, 2009 called Vortex2 across the southern and central plains. Research teams used sophisticated instruments, including satellite images. NOAA sent out a fleet of ten mobile radars and dozens other instrument trucks. They drove over fifteen thousand miles each, booked seventy-five hotel rooms each night, each housing up to 150 people at a time. They used remote-controlled airplanes. One can imagine the cost. Still, the one thing being left out of the equation is what is occurring a short distance under the ground: heat generated by the core.

The first thing scientists will point is the fact that there are tornadoes that develop in cold climates.

Consider that cold does nothing; the colder it is, the less the chances of things happening. However, add heat to any system, and things move. And sometimes, they move violently (Fig. 20).

I have a new theory on how tornadoes, earthquake, volcanoes, and hurricanes are further developed by the extraction of oil and natural gas, but I do not have time to pursue it at this time. I will submit it as a scientific paper later if I can find a source that will at least read it.

Figure No. 20: Lightning Storm

Let's Just Tear It Apart!

That is just what we are doing when oil and natural gas companies use a process called *hydraulic fracturing*, sometimes referred to as *fracking* or *splitting apart*. Oil and natural gas corporations inject liquids composed of water and chemical into wells. This action creates fractures by splitting rocks deep underground, causing oil and natural gas to escape in areas that would not be possible by standard oil drilling.

Recently, there have been many articles written on this subject. Scientists, politicians, oil and natural gas companies, and even university professors have all voiced their opinions on the subject. However, no one has been down there yet to see what is really going on. So let me put a new approach on this as to what I think is going on.

The common definition of an earthquake that is usually comprehended by most people brings to mind such places that have had catastrophic damage to lives and infrastructures, places like San Francisco in 1989, which was seen on TV all over the world. The 1989 World Series between the Oakland Athletics and the San Francisco Giants game 3 was just starting its pregame commentary when the earthquake occurred. After a brief electrical outage, ABC and CBS began continuous coverage along with the Goodyear blimp that was covering the series from overhead. It can be remembered for the destruction because it occurred during the day with live pictures, especially the ripple effect of the Central Freeway, US Highway 101 a double decker bridge that was destroyed. What I want you to notice is the ripple effect the quake had on the bridge (Fig. 21). This is the general way an earthquake acts. It's similar to the concentric circles a lightning strike makes, except an earthquake extends out more in one direction whereas lightning extends in a concentric circle. The quake occurred near the San Andreas Fault System at a depth of twelve miles. Earthquakes generally develop close to tectonic plates, where the strain between these plates are naturally occurring geological events; in this case, the North American Plate is colliding with the Pacific Plate.

Figure No. 21: Collapsed section
of Interstate 880 bridge.

There have been many tremors recently throughout Oklahoma. But there are no tectonic plates in the central part of the United States colliding with each other; as a matter of fact, Oklahoma is practically in the middle of the North American Plate. The United States Geological Survey just came out with a report that they believe there is an ancient, dormant fault line that has been reactivated by so many earthquakes in central Oklahoma. If it has been dormant for so long, what activated it?

Here is my new approach to the problem. First, have you ever noticed how many oil wells are in Oklahoma, or for that matter throughout Texas, Louisiana,

Mississippi, and all over the United States? (Fig. 22). These oil wells have been carving out canyons underground for decades. They are not earthquakes as we visualize, but are cave-ins, like what the miners experienced in Chile while digging for ore deep in the ground.

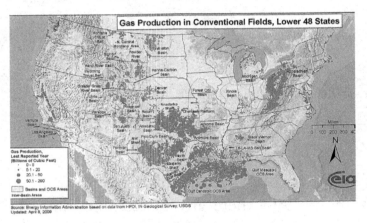

Figure No. 22: The huge quantity and location of U.S. oil wells.

There is a simple rule that states that without a foundation, the structure will collapse. Everything in life that is of any value must have a firm foundation, for without a firm foundation in our life to stabilize us, we become erratic and will fail in our efforts. The same is true with the ground we live on.

If I had continued to dig under our home when I was young, the foundation would have been compromised, and the house would not be there today. Another approach would be if a single shovel of dirt was removed every day for fifty years, do you really think the foundation would have held the house up? The same thing is happening when oil and natural gas is removed underground. The caverns left behind after oil and natural gas have been extracted are just the initial beginnings. The cave-ins are becoming much more frequent due to this new process of hydraulic fracturing. The infrastructure down under is being destroyed by these cave-ins. When water and chemicals are injected under high pressure to fracture rocks, the oil that is released weighs tons. This flow of oil released alone could cause tremors in the ground besides the falling rocks. One more thing: as the caverns get bigger and deeper, the oil and rocks falling will impact the bottom of the well with greater force, thereby creating greater tremors above ground. This is why there have been stronger recorded quakes as more and more oil and natural gas is removed. These collapses are relatively close to the surface, unlike earthquakes that occur many miles below the surface. These tremors are being felt all over the world; they are not just in the United States.

There is a possibility that if the ground is becoming unstable, which in return will become thinner, an earthquake will occur due to the strain put on tectonic plates over the whole earth. This could cause buckling and collapse of the ground in a cartographic way, whether it lies in an earthquake region or not.

Areas close to dams, rivers, lakes, and fault lines that put horizontal pressure on its surroundings will be extremely vulnerable.

Can we correct the damage that hydraulic fracturing has caused? The answer is no. Can we stop what is being done to our planet? The answer is we could, but the odds of doing so is extremely doubtful. As of the writing of this book, the price of oil and natural gas has dropped dramatically. Several oil companies have filed for bankruptcy. Now would be a good time to reexamine the process and even outlaw the procedure to protect our future safety from a more devastating destruction that will happen if this type of drilling is not stopped. Already, there are "Wizard of Oz" type of statements to protect the main foundation of oil and natural gas companies. For example, experts say these quakes are likely caused by injection wells, which are very deep wells that they are injecting wastewater and drilling byproducts into rather than the original wells drilled to extract oil and natural gas. Don't look at the man behind the curtain. Don't blame any of this on our mainstay of bringing oil and natural gas out of the ground because of fracking (Fig. 23). I believe the mounting evidence of the destructive power of hydraulic fracturing is quickly coming to the forefront, but there is a still a great threat lurking underground in the continuing extraction of oil and natural gas.

Figure No. 23: Oil field in Kern County,
CA. About 15 billion barrels of oil could be
extracted using hydraulic fracturing.

I would like to say one more thing before leaving
this subject. As damaging as hydraulic fracturing is,
we must not lose track of the loss of insulation that
oil and natural gas is providing against the extreme
heat from the interior of the Earth (Fig. 24).

Figure No. 24: Radiated heat from Earth

Now What?

The damage has been done and is being continued at an astounding rate. Geothermal energy is taking enormous amounts of heat from below the Earth's crust. Oil and natural gas are being extracted at the tune of billions of barrels and trillions of cubic feet out of the ground every year with no regard to its consequences. So what do we do now?

The first thing people look into is alternative energy. Not all alternative energy is beneficial and does not further harm our environment.

I have covered geothermal energy earlier and have shown this to be out of the question for an alternative energy source. Natural gas has been determined to be a clean form of energy. But its elimination from inside the Earth is causing extreme damage by its removal of the insulating properties. Natural gas has been protecting us from the intense heat of the core for millenniums. Along with oil, this form of energy must be stopped quickly. It will not be easy. When people's livelihood is in jeopardy, it is a daunting process. For some, it is not only their livelihood at stake; it's their entire belief of existence, much like a religion. They will fight with all means available regardless of the end result. It will take honest governments to step in and filter this out as well as scientific proof of what is really happening under our feet.

Let's begin with a few of the present new energy sources. Let's start with wind generators that tower over our landscape, ruining the scenic beauty of land and sea (Fig. 25). Environmentalist like this; why, I don't know. It seems they are so dead set on believing that the use of oil and natural gas are destroying our atmosphere that they cannot see the forest for the trees. They will try anything that they believe will not add greenhouse gases. Legitimate scientists have determined that greenhouse gases will not be a valid reason for global warming until fifty to one hundred years in the future. I contend the extraction of heat through geothermal energy, and removal of the protective properties of oil and natural gas is of immediate emergency.

Figure No. 25: Wind Mill Farm on beach
in Bangui, Central African Republic

I call wind power machines "artificial trees." Trees have been used to protect from wind, cold, and heat. This means they have been used to slow down wind or redirect wind. I suggest that wind power on top of mountains, besides being ugly, are redirecting and slowing down the speed of wind. This in turn is upsetting the natural flow of wind currents that has been going on for centuries. When these giant trees are put out to sea or across acres of land in open plains around the world, they are disrupting the natural flow of seasonal winds. This is causing the wind to be slowed down, thereby causing the natural flow to put more or less heat and cold to penetrate into areas that wind typically does not go or should go. One more thing that wind power is doing is killing large numbers of birds. If this continues over several decades, it could not only bring some birds to extinction but increase disease and harmful insects all over the world. Birds eat insects like mosquitoes and plant-eating insects that destroy crops. Some birds are used to pollinate. Other predatory birds such as eagles, hawks, and buzzards are birds that control snakes and rodents. I would not put wind generators on a list of clean-air-producing energy systems.

One type of clean air energy system I would put on a list is solar; however, I do so with reservation. Using valuable land that could grow crops does not seem to be the best usage, seeing that food and water is fast becoming our most precious commodies. Land use will be needed to be controlled due the increase in population and the decrease in land productivity because of cooler temperatures in the northern latitudes. I recommend putting solar panels on roofs, for now; later these areas will be needed to produce food (Fig. 26). Areas that could be used that are in abundance are roads and sidewalks.

Figure No. 26: Solar panels used on roof.

Nuclear energy has its problems, but it may have to be used more (Fig. 27). The problems it creates may be "necessary evils." I am one that believes science can overcome these deficiencies; like the waste product could either be reused or even sent into the sun. Operational problems are constantly being overcome and should never be relaxed. Vulnerability to a terrorist attack should be of utmost priority.

Figure No. 27: Nuclear Power Plant

Figure No. 28: Coal being excavated near surface.

Here now is the real kicker that will cause some to throw this book away. The solution to our future energy needs—before you toss it, please let me explain—is coal. Okay, now that you have not thrown the book away or blown your top and become infuriated, here is my reasoning. All throughout the book, I have stated that the removal of oil and natural gas has caused heat to rise up to surface of the Earth by eliminating the insulating qualities they provide to protect us from the extreme heat of the core. Oil and natural gas is extracted from its source deep underground where coal is near the surface (Fig. 28). I do not believe that the coal near the surface is providing any significant protection from the heat from the core any more than the crust of the Earth is containing; I had much to say about this in another part of this book. Therefore, coal should be used without any damage at this time to protecting us from the core. As for the pollution, it produces into the atmosphere, there have been many advances in recent years to control this. Some advances made recently have been capturing and storing carbon by burying it deep underground. It does not need to be stated that there are tremendous volumes of caverns available to do this. I would caution against this being a final solution to this problem. With increase in production, we should require the coal industry to invest in future advancements to further control and eliminate its effect on our atmosphere (Fig. 29). It is estimated we have more than enough of this energy source to supply the world for many years to come or until science has progressed enough to provide us with a more environmentally benign method of energy. Also limiting this base of energy to developing nations would be unfair and devastate their economies. Destroying these countries ability to a higher standard of living is counterintuitive to making our planet a better place to live. A better world means better health for all and a purpose with hope for the future.

Figure No. 29: Experimental Coal Power Plant

CPSIA information can be obtained
at www.ICGtesting.com
Printed in the USA
FSOW03n2238290716
23236FS